Y0-BZJ-323

106007957
LAKE SUPERIOR ELEM IMC
SUPERIOR, WI

591.51
WEX

106007957

Wexo, John Bonnett.

Night animals

$14.95

DATE DUE	BORROWER'S NAME	ROOM NO.

591.51
WEX

106007957

Wexo, John Bonnett.

Night animals

LAKE SUPERIOR ELEMENTARY LMC
6200 E 3RD STREET SUPERIOR WI

789839 01495 28127C 28568F   001

NIGHT ANIMALS

Published by Creative Education, 123 South Broad Street, Mankato, Minnesota 56001

Copyright ©1996 by Wildlife Education, Ltd. Copyright 1996 hardbound edition by Creative Education. All rights reserved. No part of this book may be reproduced in any form without written permission from the publisher. Printed in the United States.

Printed by permission of Wildlife Education, Ltd.

Library of Congress Cataloging-in-Publication Data

Wexo, John Bonnett.
Night animals / series created and written by John Bonnett Wexo.
p. cm.
Includes index.
Summary: Discusses the habits of animals that spend most of their lives in the dark,
such as racoons, owls, hyenas, and catfish.
ISBN 0-88682-777-9
1. Nocturnal animals—Juvenile literature. [1. Nocturnal animals.] I. Title.
QL755.5.W48   1996
591.51—dc20   95-45323   CIP AC

NIGHT ANIMALS

DISCARD

LAKE SUPERIOR
ELEMENTARY SCHOOL
SUPERIOR, WISCONSIN
106007957

Creative Education

Art Credits

Paintings by Walter Stuart

Photographic Credits

Front Cover: Rod Williams (*Bruce Coleman, Inc.*)

Pages Six and Seven: Russ Kinne (*Photo Researchers*)

Page Eight: Top, Stephen Krasemann (*DRK Photo*); **Middle,** Christopher Crowley (*Tom Stack & Associates*); **Bottom Left,** Wayne Lankinen (*DRK Photo*); **Bottom Right,** Stephen J. Krasemann (*Peter Arnold, Inc.*)

Page Nine: Top, Laura Riley (*Bruce Coleman, Inc.*); **Middle Left,** Stephen Krasemann (*DRK Photo*); **Middle Right,** Anthony Bannister (*Animals Animals*); **Bottom,** Jane Burton (*Bruce Coleman, Inc.*)

Page Ten: Tom McHugh (*Photo Researchers*)

Page Eleven: Top, Shostal Associates; **Bottom,** Kjell Sandved

Page Twelve: Runk/Schoenberger (*Grant Heilman*)

Page Thirteen: Top, John Gerlach (*Tom Stack & Associates*); **Top Right,** Richard G. Leonhardt; **Bottom,** Shostal Associates

Pages Fourteen and Fifteen: James H. Carmichael, Jr. (*Bruce Coleman, Inc.*)

Page Sixteen: Top Left, Mark Newman (*Earth Images*); **Top Right,** Leonard Lee Rue III (*After Image*); **Middle,** Mark Newman (*Earth Images*); **Bottom,** Rod Allin (*Tom Stack & Associates*)

Page Seventeen: Top Left, J.M. Burnley (*Bruce Coleman, Inc.*); **Top Right,** Rich McIntyre (*Tom Stack & Associates*); **Middle Left,** V.K. Oxon (*Animals Animals*); **Middle Right,** Shostal Associates; **Bottom Left,** Stephen J. Krasemann (*DRK Photo*); **Bottom Right,** V.K. Oxon (*Animals Animals*)

Page Eighteen: Top, H. Reinhard (*Bruce Coleman, Inc.*); **Middle Left,** Jen and Des Bartlett (*Bruce Coleman, Inc.*); **Middle Right,** Buff Corsi (*Tom Stack & Associates*); **Bottom,** Rod Allin (*Tom Stack & Associates*)

Page Nineteen: Top, George Holton (*Photo Researchers*); **Bottom Left,** John D. Luke (*West Stock*); **Bottom Right,** Tom McHugh (*Photo Researchers*)

Page Twenty: Top, J. Alsop (*Bruce Coleman, Inc.*); **Middle,** Zig Lesczcynski (*Animals Animals*); **Bottom Left,** Bertram G. Murray, Jr. (*Animals Animals*); **Bottom Right,** Stephen Dalton (*Animals Animals*)

Page Twenty-One: Top Left, Hans Reinhard (*Bruce Coleman, Inc.*); **Top Right,** Wendell Metzen (*Bruce Coleman, Inc.*); **Middle Left and Right,** Stephen Dalton (*Animals Animals*); **Bottom,** Stephen Dalton (*Animals Animals*)

Pages Twenty-Two and Twenty-Three: Nick Bergkessel (*Photo Researchers*)

Our Thanks To: Amadeo Rea (*San Diego Natural History Museum*); Ron Monroe; Lynnette Wexo

Cover Photo: Young douroucouli monkey

Contents

Night Animals are different from most of the animals that you know best. Unlike horses and cows and dogs, the night animals spend most of their lives in the dark. While you are asleep, they are hunting, eating, drinking, building nests, or caring for their young. And when you are getting up in the morning, they are just going to sleep.

Actually, this is not as strange as it may sound. At one time, many millions of years ago, *all* mammals were night animals. This was because there were dinosaurs around during the day that liked to eat mammals. And it was safer for the mammals to hide during the day and come out at night.

You might say that the dinosaurs and the mammals *shared* the world. During the day, the dinosaurs could take all the food and water they wanted. And at night, the mammals could take all that they wanted. By sharing in this way, the two groups of animals left room in the world for each other.

This kind of sharing is still going on today. Except that the dinosaurs are gone, and many mammals have become daytime animals. During the day, some of the animals in the world take what they need—and after dark, other animals take what they need. And just as it did millions of years ago, this sharing makes room for more animals.

Daytime animals, by the way, are called *diurnal* (DIE-URN-UL). And the night-time creatures are called *nocturnal* (KNOCK-TURN-UL) animals. There is even an "in-between" group that is active between the night and the day, during the dawn hours and at dusk. These are called *crepuscular* (KREP-US-CUE-LUR) animals.

In many ways, life at night is not much different from life in the daytime. There are still predators that have to hunt, and prey that must hide. But to do these things in the dark, nocturnal animals have to be different from diurnal animals in some ways. For example, owls have to have eyes that can see better in the dark. And foxes must have noses that can smell better than most daytime animals. As you will discover, there are many creatures that have fascinating ways to live as night animals.

Gerbils are very alert at night. These nocturnal rodents listen carefully for the sounds of foxes and other night hunters. If one of them hears something, it stomps the ground to warn the others. Then it quickly scampers away to look for a place to hide.

The different times of day and night offer different advantages (and disadvantages) to different animals. Since people are most active during the day, we tend to think that daytime is the best of all times. But this is not really true.

Some animals would not survive very long in the bright light of day. They might be more easily caught by predators. Or their bodies might get too hot and dry up. Or they might not be able to compete with other animals for food.

If they could talk, many animals would probably tell you that being crepuscular and nocturnal is every bit as good as being diurnal.

Most birds are diurnal, for obvious reasons. When they fly, they need the daylight to help them see where they are going. Eagles and most birds of prey often soar high in the sky to look for prey. This would be impossible at night.

BALD EAGLE
(DIURNAL)

SNOW LEOPARD
(CREPUSCULAR)

Many predators are crepuscular because their prey is easiest to catch at dawn and dusk. The best time for a snow leopard to catch a mountain goat is at twilight, so the snow leopard is crepuscular.

CHIMPANZEE
(DIURNAL)

GRIZZLY BEAR
(CREPUSCULAR)

Daylight makes it easier for plant-eating animals to find food. But it also makes it easier for predators to find the plant eaters —so daytime can be a very dangerous time.

The heat of the day is uncomfortable for some animals, so they prefer the coolness of the dawn hours or the late evening. Bears are a good example of this. During the hot daytime hours, they usually find a shady place to take a nap.

EASTERN CHIPMUNK
(DIURNAL)

Safety is one of the main reasons why animals are nocturnal. It is easier to hide in the dark, and harder for predators to see you. This raccoon certainly has a better chance of robbing a trash can when nobody can see him.

RACCOON
(NOCTURNAL)

MULE DEER
(CREPUSCULAR)

Some animals may be nocturnal *and* diurnal, depending on the circumstances. Most of the time, jackals are daytime animals. But in places where they feel threatened by people, the timid jackals become night creatures. They live and hunt in the safety of darkness.

BLACK-BACKED JACKAL
(NOCTURNAL AND DIURNAL)

RED FOXES
(NOCTURNAL)

Some small animals have an easier time finding food at night than they would during the day. Foxes and other small predators don't have to compete with as many large animals for food as they would in the daytime.

Many night animals take over "jobs" that are done by other animals during the day. For example, owls hunt many of the same kinds of creatures that eagles and hawks hunt during the day.

TAWNY OWL
(NOCTURNAL)

MARGAY
(NOCTURNAL)

9

Very sharp senses are absolutely necessary for most night animals. To find food in the dark, to escape from danger, or just to find their way back home, they must have senses that are incredibly good. In fact, very few daytime animals can match nocturnal animals when it comes to seeing, hearing, feeling, and touching.

FINGERS
THAT CAN SI

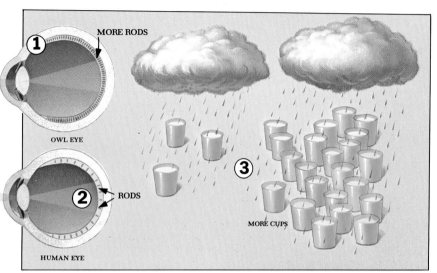

MORE RODS

① OWL EYE

② RODS

HUMAN EYE

③ MORE CUPS

Owls and many other night animals have special eyes for seeing better at night. These eyes are similar to yours. Like your eyes, they have a *lens* ① that gathers light. And they have many tiny cells at the back of the eye that are called *rods* ②. The rods are sensitive to light. And the more rods there are in an eye, the more sensitive the eye is to light. You could compare the rods to cups that you put out in the rain ③. If you only have a few cups, you don't catch much water. If you have many cups, you catch a lot. There is only a little light at night, but the owl's eye catches more of it. So the owl can see better.

SPECIAL EYES

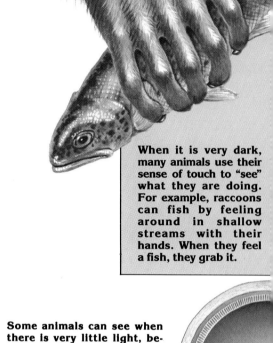

When it is very dark, many animals use their sense of touch to "see" what they are doing. For example, raccoons can fish by feeling around in shallow streams with their hands. When they feel a fish, they grab it.

Some animals can see when there is very little light, because their eyes can see every ray of light *twice*. First, the light comes into the eye and strikes the rods Ⓐ. Then it bounces off a kind of mirror in the back of the eye Ⓑ and hits the rods again. The "mirror" is called a tapetum (tah-PEA-tum).

Ⓐ
Ⓑ

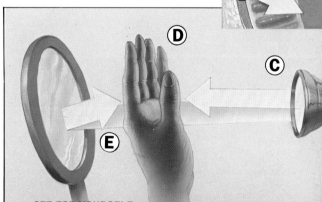

Ⓓ
Ⓒ
Ⓔ

SEE FOR YOURSELF how a tapetum doubles the light tha hits the rods. Pretend your fingers are rods. Spread you fingers in front of a mirror and shine a light on them Ⓒ See how the light hits the fingers once Ⓓ, and how hits them again Ⓔ after bouncing off the mirror.

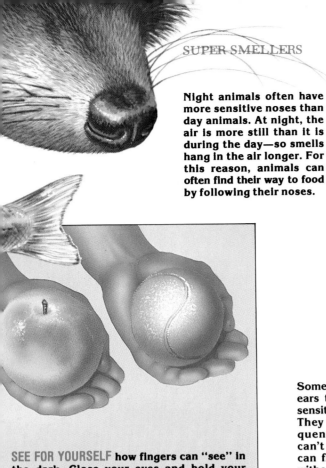

SUPER SMELLERS

Night animals often have more sensitive noses than day animals. At night, the air is more still than it is during the day—so smells hang in the air longer. For this reason, animals can often find their way to food by following their noses.

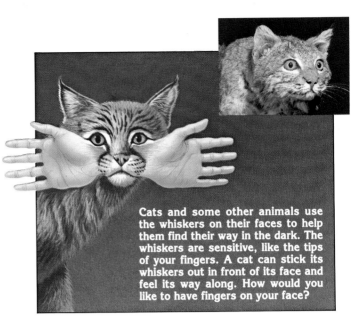

Cats and some other animals use the whiskers on their faces to help them find their way in the dark. The whiskers are sensitive, like the tips of your fingers. A cat can stick its whiskers out in front of its face and feel its way along. How would you like to have fingers on your face?

SEE FOR YOURSELF how fingers can "see" in the dark. Close your eyes and hold your hands out. Have a friend put a peach in one hand and a tennis ball in the other. Both of these things are round and about the same size. But just like a raccoon, you can tell which one is good to eat and which one is not.

Some night animals have ears that are much more sensitive than human ears. They can hear high frequency sounds that we can't hear. In fact, bats can fly in total darkness without bumping into anything—because they can *hear* the things that are around them.

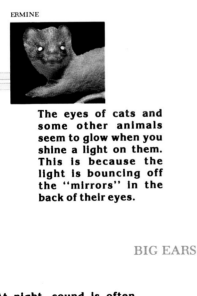

ERMINE

The eyes of cats and some other animals seem to glow when you shine a light on them. This is because the light is bouncing off the "mirrors" in the back of their eyes.

BIG EARS

At night, sound is often the best way for animals to find out where other animals are. The air is usually still, and sounds may carry a long way. By listening, a predator may find out where its prey is. Or the prey may hear the predator coming, and escape. To hear sounds better, many night animals have big ears.

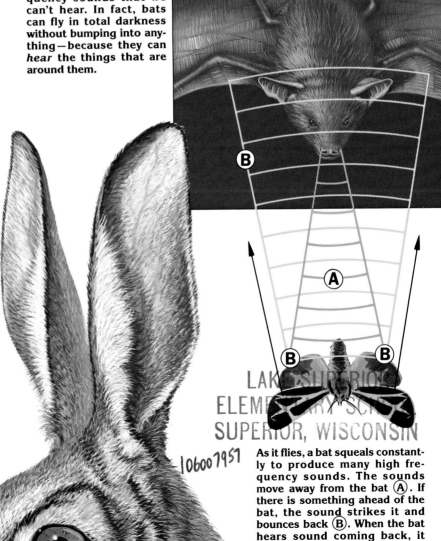

LAKE SUPERIOR ELEMENTARY SCHOOL SUPERIOR, WISCONSIN
106007951

As it flies, a bat squeals constantly to produce many high frequency sounds. The sounds move away from the bat (A). If there is something ahead of the bat, the sound strikes it and bounces back (B). When the bat hears sound coming back, it knows there is something ahead. This is called echolocation (EK-oh-low-KAY-shun).

11

Hunters and their prey

Hunters and their prey carry on a never-ending battle at night. Sometimes the hunter wins, but most often the prey gets away. This is a good natural plan, because if too many animals were caught, the prey would soon die out. And if that happened, the hunters would die out too.

At night, the hunters must have special methods for hunting. They need extra-sharp senses to find their prey in the dark. And they must be very quiet, for the animals they hunt also have sharp senses, and special ways of escaping at night.

Being quiet is especia[lly] important when hunt[ing] at night, because ma[ny] prey animals have go[od] ears. So cats have hea[vy] fur and soft pads on t[he] bottoms of their fe[et.] This helps them wa[lk] quietly, so they can g[et] close to their prey wit[h]out being heard.

Special defenses are sometimes needed to combat special hunting skills. For example, many moths have a good defense against the echolocation used by bats. They have highly sensitive ears, and when they hear a bat's high-pitched sounds, they dive for the ground.

This nocturnal hunter looks like it has whiskers. That's why it is called a catfish. But its "whiskers" are really *barbels* (rhymes with *marbles*), and they are flesh instead of hair. The catfish uses its barbels to feel and *smell* its way along the river bottom.

CHANNEL CATFISH

Most hunters will take advantage of any opportunity to find prey. But few are as shrewd as foxes. Sometimes a fox will even follow a badger, because the badger will lead it to a good patch of earthworms, which foxes like to eat.

Often, in contests between nocturnal hunters and their prey, the winner is the one that hears the other first. Even though a bobcat has good ears, and usually hunts very quietly, an alert jackrabbit will usually hear it in time to get away.

WHAT AN OWL SEES

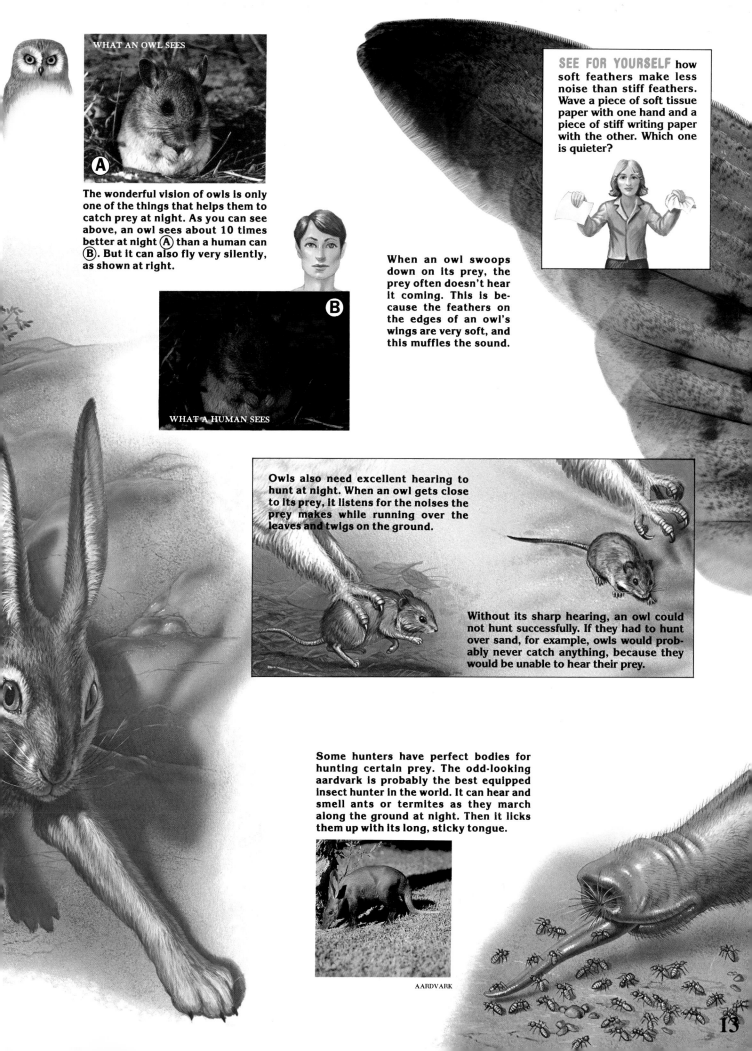

A

The wonderful vision of owls is only one of the things that helps them to catch prey at night. As you can see above, an owl sees about 10 times better at night (A) than a human can (B). But it can also fly very silently, as shown at right.

B

WHAT A HUMAN SEES

SEE FOR YOURSELF how soft feathers make less noise than stiff feathers. Wave a piece of soft tissue paper with one hand and a piece of stiff writing paper with the other. Which one is quieter?

When an owl swoops down on its prey, the prey often doesn't hear it coming. This is because the feathers on the edges of an owl's wings are very soft, and this muffles the sound.

Owls also need excellent hearing to hunt at night. When an owl gets close to its prey, it listens for the noises the prey makes while running over the leaves and twigs on the ground.

Without its sharp hearing, an owl could not hunt successfully. If they had to hunt over sand, for example, owls would probably never catch anything, because they would be unable to hear their prey.

Some hunters have perfect bodies for hunting certain prey. The odd-looking aardvark is probably the best equipped insect hunter in the world. It can hear and smell ants or termites as they march along the ground at night. Then it licks them up with its long, sticky tongue.

AARDVARK

13

Eyes this big see many things. This Great Horned Owl keeps a sharp watch for insects and rodents which will be its meal.

A safe place to hide is what most nocturnal animals look for in the daytime. Many find shelter in trees and rocks. Others dig or build their own shelters. And some nocturnal animals get added protection from *camouflage*. In other words, their colors blend with their surroundings.

Although they try to stay hidden, it is sometimes possible to see night animals in the daytime. They do not *always* sleep during the day. When disturbed by predators, they must be ready to run or defend themselves. And if they are hungry enough, many of them will even look for food.

RACCOON

SNOWSHOE HARE (SUMMER)

This hare is almost as hard to see in the daytime as it is at night. In spring and summer, its coat is brown. It blends in well with the weeds and grass.

In late autumn, the snowshoe hare grows a new coat of warm winter fur. And this fur is white, so it can't be easily seen in the snow.

SNOWSHOE HARE (WINTER)

Trees are favorite hiding places for many nocturnal animals. Raccoons often spend the daytime in hollow trees. And in the winter, they may stay inside their tree homes day and night until the weather turns warm again.

WISCONSIN BADGER

Digging a hole is a very common way to build a shelter. But badger holes are something special. Badgers dig large, comfortable dens that may be used for more than 100 years — not by the same badger, of course! Usually, six or seven badgers share a single den. And one of these dens may stretch 65 or 70 feet (20 meters) into a hillside.

16

Since night animals like the dark, many of them spend the day in dark places. Most bats live in caves, where they are safe from predators—and from the sunlight.

RED FOX

LITTLE BROWN BAT

Night animals do not like to hunt in the daytime. But if they get hungry enough—or if they have babies to feed—many of them will anyway. Foxes are among the most successful nocturnal animals in the world because they adjust well to harsh conditions.

There are many different species of owls, and they have different nesting habits. Some nest in tree hollows, others in old abandoned buildings, and a few keep their nests on the ground. But most owls build nests in trees, where their colors make them hard to see.

GREAT-HORNED OWL

HEDGEHOG

Some nocturnal animals are so well camouflaged that they don't need a shelter. The hedgehog often builds its little nest out in the open. Can you find one sleeping in its nest above Ⓐ? But the hedgehog does have another defense it can use if it needs to. It can roll itself into a tight little ball Ⓑ. When it is all rolled up, only the sharp spikes on its back will show Ⓒ. This quickly discourages most predators.

Night animals do surprising things.

For a long time, people didn't know very much about the habits of nocturnal animals. They are not as easy to study as diurnal, or even crepuscular, animals. But in recent years, scientists have designed better cameras and other equipment for studying animals at night. And their studies have led to some wonderful surprises.

"Small but mighty" is a good way to describe weasels. They are among the smallest of all hunters, but they make up for their size with amazing speed and agility. The weasel usually hunts for mice and voles, dashing over the ground by leaps and bounds. And with its skinny body, it can even reach deep inside a mouse's hole.

COMMON WEASEL

SPOTTED HYENA

People used to think that all hyenas were scavengers (animals that eat the meat left behind by predators). But we now know that some hyenas are true hunters—and very good ones at that. Spotted hyenas hunt together in large packs, and they can tackle even the largest antelopes on the African plains.

BLACK BEAVER

Beavers usually work the night shift. And if you have ever seen a beaver dam or lodge, you know what good workers they are. But amazingly enough, they do most of their work at night, "by ear." They figure out where to place their logs by listening to the flow of the water.

This odd-looking bird is full of surprises. The kiwi is found only in New Zealand, where it lives on the ground, and only comes out at night. It has tiny wings and cannot fly. On its face it has long whiskers. And its tube-shaped bill has a pair of nostrils at the very tip. At night, the kiwi finds earthworms by poking its bill into the soil and smelling them.

KIWI

This excellent night hunter has the face of a raccoon, the body of a leopard, and the long tail of a monkey. But the genet is related to civets and mongooses. Mostly, genets live in Africa, but some also live in the Middle East and in southwestern Europe.

DOUROUCOULI

The slow-moving loris is safer at night. One surprising thing about a loris is the incredible strength in its fingers. It can hang from a tree branch for hours. And it clings so tightly that it's practically impossible to pull loose.

SLOW LORIS

The only truly nocturnal monkey in the world is the South American douroucouli, or owl monkey. It spends most of its life in the trees, where it manages to find all of its food. And because it is nocturnal, it gets all the water it needs from the dew.

Flying at night can be dangerous. Even with excellent nocturnal vision, it is hard to see very far in the dark. So the night sky has fewer flying creatures than the daytime sky. Most birds that do fly at night are limited to short-distance flights. And in order to see where they're going, they do not fly very high off the ground.

However, the night sky *is* full of surprises. Many unusual animals take to the skies at night. In fact, the best night fliers probably aren't even birds—they are bats.

One of the more unusual animals of the night sky is the flying squirrel. During the day, it would become an easy meal for a hawk or an eagle. So flying squirrels are only active at night. And they don't actually *fly*— they *glide*. By stretching out the loose skin between their front and rear legs, they can glide from treetop to treetop. They have been known to travel as far as 100 yards (91 meters) through the air.

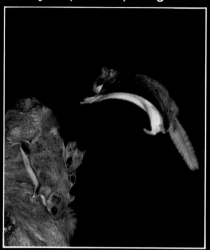

FLYING SQUIRREL

FLYING GECKO

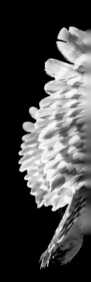

Did you know that there are even lizards that can "fly"? There are 16 kinds of them that can float through the night sky. Actually, they don't fly like a bird. Instead, they use flaps of skin on their bodies like wings, and *glide* from place to place.

OILBIRDS

These beautiful creatures are probably the best night flyers of all the birds. In fact, they can fly almost as well as bats, and their lives are very similar to the lives of bats. During the day, they live in caves. And when they fly, they use the echoes of their screams to guide them.

Owls are excellent night hunters, but they usually hunt close to the ground. They do not fly great distances, but they have a re-markable ability to fly through heavily wooded areas without bumping into trees.

WOODCOCK

At sunset, when most herons are flying to their nests, this night heron is just waking up. During the night, it prowls the shallow waters where other herons feed during the day. And this nocturnal lifestyle has been very good for the herons. In terms of numbers, this species is one of the most successful herons in the world.

YELLOW-CROWNED NIGHT HERON

The woodcock is one of the few birds that is a really good nighttime flyer. It can fly at high speeds. And just when you think it is going to crash into something, it suddenly pulls up into the air.

LITTLE OWL

Because they can "see" in a different way, bats can fly higher, farther, and faster at night than most nocturnal birds. And they are incredibly good at catching prey. The fishing bat shown below uses its echolocation system to detect fish that are swimming near the surface of the water. Then it swoops down and uses its long claws to grab the fish.

SOUTHERN FLYING SQUIRREL

Index